EXPLORING CITIES
BEDTIME RHYMES

Written and Created by
Dr. Jonathan Reichental
& Brett Hoffstadt

This book belongs to:

First Print Edition December 2021

Cover design by Safeer Ahmed

Copyright © 2021
Dr. Jonathan Reichental
& Brett Hoffstadt
All Rights Reserved

Paperback ISBN: 978-1-7376099-2-6
eBook ISBN: 978-1-7376099-1-9

www.smartcitybook.com/kids

This book is a product of
Innovation Fountain
www.innovationfountain.com

Made with love in California, USA

Dedications

I dedicate this book to my mother, Evanne. My brothers and I called her "Mammy." She was my world and I think about her every single day.

                        Dr. Jonathan Reichental, December 2021

I dedicate this book to my mother who made bedtime rhymes and stories - with occasional assistance from my father – a formative part of my early childhood. I hope this book provides fond enjoyment to nurture many future lives.

                        Brett Hoffstadt, December 2021

Contents

Ready for an Adventure?   1

History of Cities   3

Cities Today   13

Future of Cities   25

Glossary   45

About the Authors   51

Acknowledgements   53

Additional Resources   55

Ready for an Adventure?

Congratulations, Explorer!

Together, we are about to embark on an incredible journey to discover the past, present, and future of cities.

Today, more than half of the people in the world live in cities. In the years ahead, many more people will join them. Our future belongs to cities!

A healthier and happier future in cities will require us to create new ways to get around, use clean **energy**, manage **waste**, create a better environment, ensure inclusion for all, and do much more.

Creating and adopting new ideas that improve quality of life in our cities is what the future needs to be about. Cities are important to all of us.

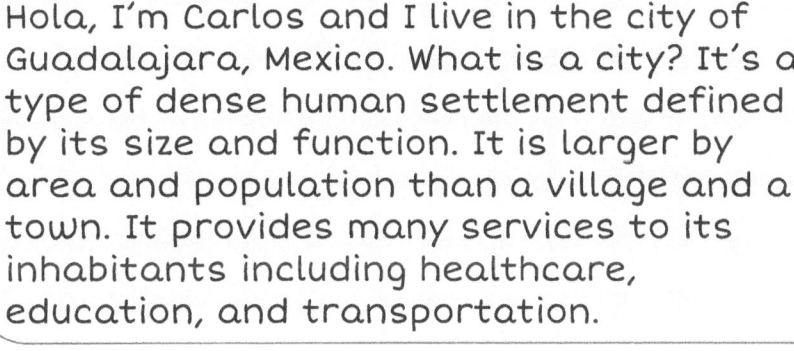

Hola, I'm Carlos and I live in the city of Guadalajara, Mexico. What is a city? It's a type of dense human settlement defined by its size and function. It is larger by area and population than a village and a town. It provides many services to its inhabitants including healthcare, education, and transportation.

As you read the rhymes, whenever you see a word in bold, you'll find a definition of it in the glossary at the back of the book.

If you enjoy learning about cities, you might consider this book's companion edition, *Exploring Smart Cities Activity Book for Kids*, which includes games, puzzles, coloring, and more.

We live in exciting times! Ensuring cities provide a great future for all of us is important and meaningful work.

Perhaps the next great city leader is holding this book right now. Okay, Explorer, it's time to launch our journey of discovery.

Let's go!

History of Cities

"Namaste, I'm Bhavna and I live in the city of Bangalore in India. Let's explore the history of cities. Understanding the past can help us prepare for the future."

"Konnichiwa, I'm Akio. I currently live in the city of Yokohama in Japan. In this section we'll look at why cities were created."

Before There Were Cities

Once we were wanderers who moved from place to place,
searching for food and trying to stay safe.

There were no cities built yet to be found,
our impact was little more than footprints on the ground.

But things always change, we know this to be true;
we discovered ways to live that were completely new!

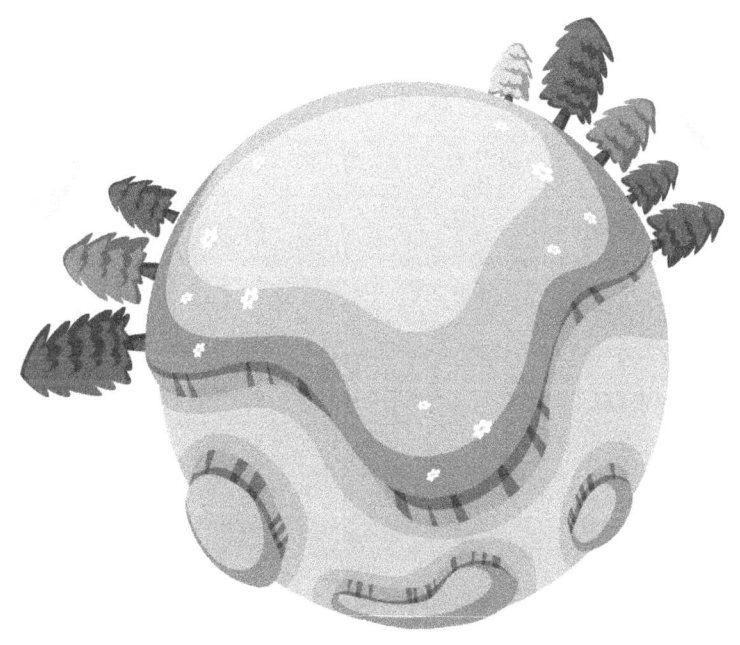

The first cities began to emerge
10,000 - 20,000 years ago.

People Created Settlements

We learned to be farmers and grow things to eat,

settling in one place,

to grow fruit, vegetables, and to raise meat.

With plentiful food we lost one big worry;

we no longer had to pack up and move in a hurry!

> **Agriculture** is another word for farming. It includes growing and harvesting crops and raising animals.

Settlements Grew To Be Cities

Cities are different from villages and towns.
You're unlikely to hear farm animal sounds.

Cities have many more people and take up
lots of space.
Look out the window - is that your neighbor's face?!

Oí, I'm Lucas from Goiânia, a city in Brazil. What are some of the differences between living in **urban** and **rural** areas?

What do we mean by the words **urban**, **suburban**, and **rural**?

- ➢ **Urban** areas are the dense, central areas of a city.
- ➢ **Suburban** areas are usually homes that surround an **urban** area.
- ➢ **Rural** are the countryside areas outside cities.

Cities Arrive On The Scene

If you live in a city, what is its name?
Does it have some unique claim to fame?

If you went to another city, where would it be?
Talk about the things there that you'd like to see.

Cities Become Popular

With cities becoming ever more popular,
people began arriving from near and far.

Many new shops opened in neighborhoods.
New arrivals brought diverse services and goods.

At first there was barter, the act of swapping.
Later came money which was easier for shopping.

With the hustle and bustle each and every day,
the appeal of cities was here to stay.

Yassou, I'm Elias. I live in the city of Patras in the wonderful country of Greece. Cities became popular because there were great jobs available for all types of skills. Cities also provided a larger variety of goods and services. What other reasons do you think made cities so popular?

Cities Created Problems

Conditions could be unhealthy in early places,
and this put a frown on too many faces.

Some of the streets were awfully smelly,
and made many people sick in the belly.

Too many crooks were engaged in crime,
this was unpleasant and lasted for some time.

An important development in cities was the introduction of **sanitation**. This meant providing cleaner water, removing of trash, and using toilets.

In Time, City Life Improved

Sanitation and **medicine** reduced diseases,
by lowering infections and the number of sneezes.

After a while, improvements did succeed,
cities got cleaner, healthier, and safer indeed.

With better living being the prize,
cities continued to grow in size.

Cities Today

Marhaba, I'm Atif from the city of Riyadh in Saudi Arabia. Today, cities are remarkable places, but they also have problems we must solve together. We'll look at some of those in this section.

Annyeong, I'm Duri. I live in the city of Busan in South Korea. I love my city and I want to help improve it too. Let's explore areas where cities need help.

Understanding The Built Environment

Cities have many **structures** such as schools.
They also have offices, bridges, and public pools.

What types of **structures** are on your street?
How about places where your friends might meet?

The **built environment** (or **built world**) is the **structures** that support where people live and work.

It's important that our cities are built so that everyone can access all facilities and can move about freely. When cities are accessible and everyone is valued and included in opportunities, this is called being inclusive.

Smart cities are inclusive cities!

Cities Are Full Of Diverse Buildings

Buildings are big and many are small.
Imagine if you could explore them all!

Some are famous and some are not.
One thing is certain though, there sure are a lot!

Water Is Essential For Life and Cities

Water is essential for us to all exist,
from drinking to cleaning it forms a long list.

Many water pipes are really old.
They struggle to move liquids that are hot and that are cold.

Old pipes can create **waste** from leaky holes.
A smart city will fix them as one of its **goals**.

Hola. I'm Miguel. Carlos is my big brother. I hope I don't waste too much water. What ideas do you have that can help to conserve water?

Cities Need Lots Of Energy

It's a fact that cities need plenty of power, for running the streetlights, the internet, and your shower.

In the past **fossil fuels** have made things run. We're getting smarter now using more **wind** and sun.

In addition to the wind and sun, we can create clean energy through the motion of waves, and also from heat deep underground.

As of 2020, **solar** is the cheapest form of **energy** in history! Why do you think that is?

Lots Of Ways To Get Around

People and goods must get around.
There are all types of transports that can be found.

There are trucks and trains, bikes and cars,
vans and planes... eventually rockets to the stars?

Let's not forget there's walking on the street,
it's healthy and fun to get around by your feet.

Rockets will be required to travel to cities on the Moon and Mars. Would you travel on a rocket?

Public Safety Keeps Us Safe

Police, fire, and paramedics too.
They are each a city **public safety** crew.

Sometimes there are fires that simply won't quit.
Call the fire service to put them out - lickety-split!

Police enforce the law on the street.
This job is called being on the beat.

Ambulances fetch a patient quick,
to help a person that may be sick.

Fighting fires, helping the sick, and solving crimes,
we rely on **public safety** through
these trying times.

For kids, **public safety** people are sometimes known as **community helpers**.

Healthcare And Medicine

Cities have **hospitals** to care for the sick.
In a crisis they'll send an ambulance quick.

Great **healthcare** is what communities need.
It's an essential ingredient for cities to succeed.

Healthy cities are smart cities.

These cities prioritize projects that improve the physical and mental well-being of their community members. These cities also include **preventative** health activities.

Public Works Manage City Projects

Cities can't avoid some wear and tear.
Public works crews always have stuff to repair.

They must look too at things to improve,
for a city to get smarter and find its future groove.

Public works are government projects for building things like roads, schools, and bridges.

Communication Infrastructure Connects People And Things In Cities

Messages from phones travel over air and underground.
*That's where data **communications** can be found.*

Infrastructure *is needed too for the Internet and TV.*
Without it all around us there'd be nothing to tune into see.

Hi, I'm Olivia from the city of Kitchener in Canada. Today, access to the Internet is as important as electricity. But many people are still not connected. To ensure that there is equal access to the Internet, smart cities focus on connecting everyone.

Future of Cities

Building A Better Future

*Quality of life requires **innovation**,*
to make cities great in every nation.

With lots of new ideas in the mix,
there are plenty of things a city can fix.

Kóyo, I'm Bimpe. I live in the city of Porto-Novo in Benin. I love to come up with new ideas. What ways do you think of new ideas?

Smart Cities Need Smart Ideas

Urban innovators have their hands full, trying to make cities smarter and more **sustainable**.

With bold ideas and **technologies**, cities can achieve many new possibilities.

> **Urban innovation** means creating ideas and solutions to help to make cities run smarter.

Technology To The Rescue

Sensors, smartphones, and various computers, these are **technologies** that are defining our futures.

Technology can be good or bad, depends on where you sit.
In our smart city future let's make sure it's a benefit.

Data, which are facts, statistics, and units of information, are really useful in cities. Data can be used for better decision-making, **innovation,** and building trust.

Sometimes, criminals will try to break city computers to steal things or cause other problems. **Cybersecurity** uses special tools to help prevent these bad things from happening.

Connecting Our Cities With The Internet of Things

Connecting to the Internet brings more than just fun.

It can help a smart city get its services done.

For example, the **Internet of Things** or IoT, can help find out if a parking space is free.

From water leaks to **security** protection, IoT **sensors** can help with detection.

The **Internet of Things**, also known as IoT, is a Network of connected devices and people that collect and share information.

Transportation Everywhere

Often it's better to walk than use a car,
especially when the distance isn't that far.

For transport, it's good to have a choice of more,
to work, to shop, or to just go and explore!

Hi, I'm Melina. I live in San Francisco in the United States. I love my skateboard. It's a fun way for me to get around.

Bicycles Rock And Roll!

How often do you ride a bike?
It's healthy and fun, what's not to like?

Bicycles in smart cities are all the rage.
Not just for kids but for every age!

Whoosh! There Goes The Train

It's great to live near a train station.
You can walk and ride to your destination.

Trains can be fast and some may be slow,
but they're still often a better way to go.

Why do you think it may sometimes be better to use a train than a car?

Drones In Our Cities

Drones will drive the streets and fly up high. They'll go beep-beep to avoid you and buzz through the sky.

Deliveries, inspections, and more without a crew. What do you think **drones** could do?

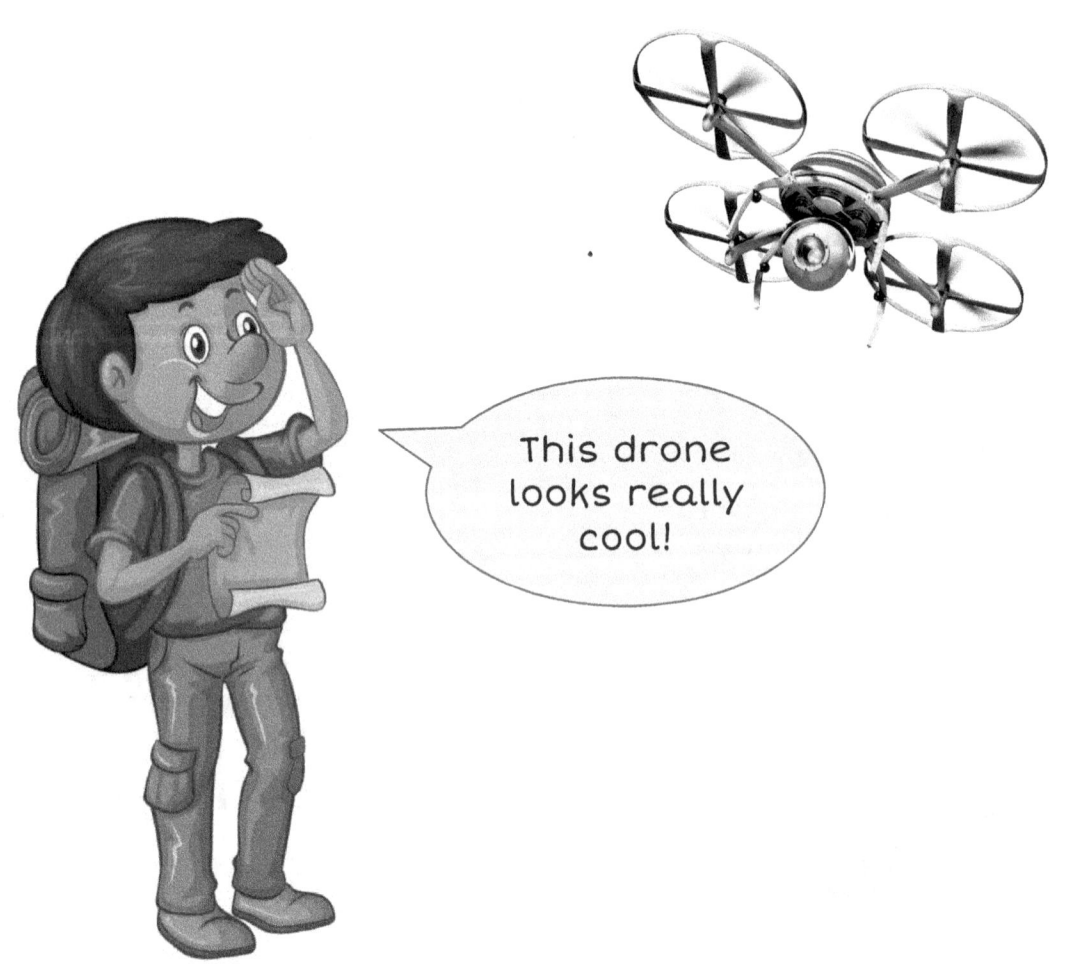

This drone looks really cool!

Fly In A Car In The Sky

Soon it may be common for cars to fly.
You'll find yourself a passenger in the sky!

They'll take all forms of shapes and sizes.
Some will even compete in races for prizes.

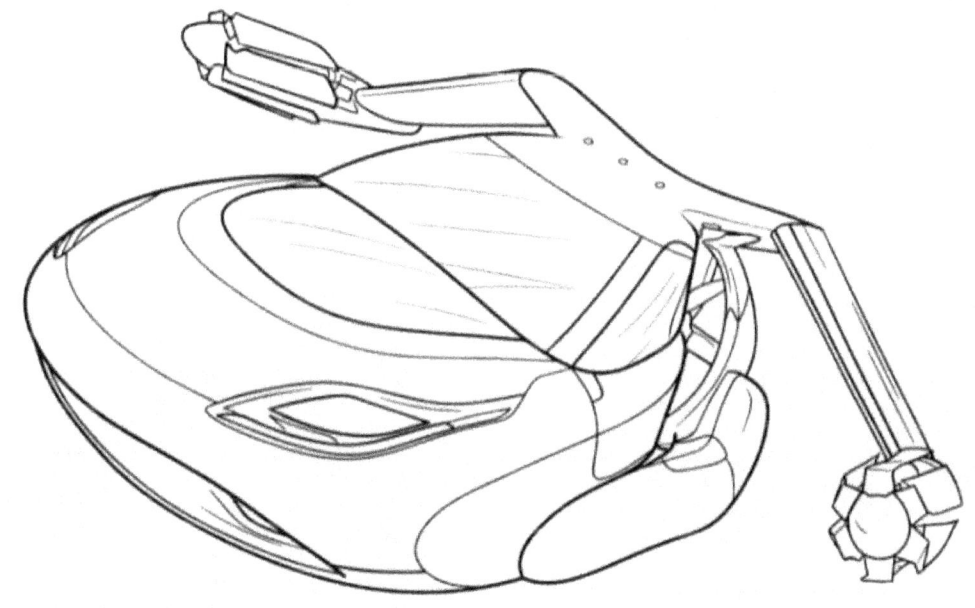

Imagine The Future Of Transportation

Salut, I'm Hugo. I live in the great city of Marseille in the South of France. The way that we travel is changing all the time. What are your ideas about how we might get from place to place in the future?

Cities Must Bring Power To The People

*Burning fuels like **oil**, **gas**, and **coal**,*

have powered cities in their historical role.

*Alternative sources of **energy** from the **wind** and the sun,*

are clean and abundant because of where they come from.

Certain human activities such as burning **oil** and **coal**, causes a **gas** called carbon dioxide to be released into the atmosphere. These are called carbon emissions. Scientists believe carbon emissions are contributing towards climate change. Smart cities will **reduce** these emissions.

Sustainable Is Attainable In Cities

Using renewable resources as **energy**,
for our future that's a better strategy.

These non-carbon sources are a lot more clean.
With zero emissions, that's why they're called green.

Less familiar energy sources

Hydro: Flowing water
Biomass: Plant & animal material
Nuclear: Splitting atoms
Solar: The Sun
Turbines: Wind
Geothermal: Heat from inside Earth

Where Does Our Waste Go?

Cities create a lot of garbage such as plastic.
Smarter ways to manage this would be fantastic.

When we throw away our **waste** it doesn't disappear.
Where it goes, what harm it does, is often unclear.

The Great Pacific Garbage Patch is an ocean area full of plastic that's three times the size of France and located between Hawaii and California.

To Be Smart
Reduce, Reuse, And Recycle

Creating **waste** without a plan can no longer be excused.

Junk can be reduced, recycled, or reused!

Let's get smarter about our trash,

or else we'll continue to see a pollution backlash.

What do each of these words mean?
Reduce: Creating less **waste** in the first place.
Reuse: Using something more than once.
Recycle: Process to convert **waste** into new things.

A Blueprint For Peace And Prosperity

Cities can become more **sustainable**.
Our progress proves that this is attainable.

Reducing consumption is really smart.
Our future will be brighter when we each do our part.

> The 17 **Sustainable** Development **Goals**, also known as the SDGs, were created by the United Nations in 2015 as a **blueprint** to achieve a better and more **sustainable** future for all by 2030.

Building A Better World Together

City success depends on what we do.
It may mean out with old ideas and in with something new.

We can build our cities to be smart.
There's a role for every one of us to do our part!

As a cyclist I'm happy that more cities are creating safe places for me to ride my bike.

Together We Can Create Smarter Cities

Big ideas and bold ones too,
we need suggestions from smart people like you!

Let's all build our cities as amazing places,
filled with joy, love, and meaningful spaces.

> Every one of us has the power to make the world a little better every day. What will you do tomorrow to show some kindness, make someone smile, or help your family, friends, or city?

Glossary

A

Agriculture	Growing and harvesting crops and raising animals

B

Barber	Person who cuts and styles hair
Biomass	Plant or animal material used as fuel to produce electricity or heat
Blacksmith	Person who makes things from steel and iron
Blueprint	Drawing that shows people how to build something
Built Environment	Structures that support where people live and work
Bullet Train	High-speed passenger train
Butcher	Person who carves and sells meat

C

Carpenter	Person who builds things out of wood
City hall	Building for city workers
Coal	Fuel that is the result of plants being turned into fossils
Cobbler	Person who makes and repairs shoes
Collaboration	Working together with others for a common goal
Communication	Sending messages between things or people
Community Helpers	Person who helps with the well-being of others in the community
Courthouse	Building that contains courts for holding law hearings and jury trials
Cybersecurity	Protecting computers from criminals

D

District	Section of a city
Diversity	Things or or a group of people with variety
Doctor	Person who treats sick or injured people
Downtown	Central area of a city with a lot of business activity
Drone	Unpiloted ground, water, or air vehicle

E

Economics	Subject of how goods and services are produced and consumed
Emergency	Unexpected situation that presents serious risks
Energy	Power to do work
Engineering	Applying science, problem solving, and creativity to design, build, and operate things
Experiment	Activity without a known result performed in order to learn something

F

Fire Station	Building that contains and supports fire fighters and their equipment such as fire trucks
Flying Car	Vehicle that can operate on roads as well as in the air
Fossil Fuel	Fuel that comes from the ancient remains of plants or animals. Examples are oil, gas, and coal.

G

Gas	Fossil fuel that is in gas form
Geothermal	Energy that uses temperature differences (thermal) in the ground (geo) to generate power
Glassmaker	Person who makes objects out of glass
Goal	Desired result of an effort
Grocer	Person who operates a store that sells food

H

Hackathon	Event where people collaborate to quickly create a basic solution
Healthcare	Services to prevent and treat sick people
Hospital	Building where doctors, nurses, and healthcare workers provide healthcare
Hydro	Use of water such as hydropower which is energy from the motion of water
Hydrogen	Light and abundant element of nature. The first element on the periodic table

I

Inclusion	Taking care to include people who are typically not included
Infrastructure	Basic physical items in a city such as bridges, roads, poles, and pipes
Innovation	Idea that is made into a new product or service
Integration	In solutions this is the process of uniting different things
Internet of Things (IoT)	Network of connected devices and people that collect and share information

J

Jail	Building where people who commit crimes are held as punishment

L

Lab	Place for creating experiments
Livability	The degree to which conditions produce a better quality of life

M

Mayor	Person elected to help run a city
Medicine	Practice of dealing with disease or injury
Megacity	City with at least 10 million people
Midwife	Person who assists with childbirth
Municipality	Another word for a type of city

N

Nomadic	Act of wandering and not living in any permanent place. These people are called Nomads.
Nuclear	Type of energy created using atoms
Nurse	Medical worker who takes care of the sick and injured and also assists doctors

O

Oil	Fossil fuel mainly composed of plants that have turned into liquid over a long period of time

P

Pilot	Trying an experiment first to see whether it makes sense to build something
Planning	Thinking of all the activities needed to reach a goal
Police	People responsible for enforcing laws, arresting people, and preserving peace in a community.
Police station	Building where police work when they are not out in the community
Post Office	Building where mail and packages are processed for delivery
Potter	Person who builds and sells items made of clay
Preventative	Protecting or guarding against serious harm, damage, or injury; being proactive
Prototype	Version of a solution that is tested with customers
Public Safety	Government work dedicated to the health, safety, and security of communities
Public Works	Government projects for building things like roads, schools, and bridges

R

Recreation	Activity that provides fun and relaxation from work
Recycling	Process to convert waste into new things
Reduce	Creating less waste in the first place
Refinery	Facility that processes oil so that it can be used for energy
Reuse	Using something more than once
Rural	Countryside area away from a city

S

Sanitation	Products and services for keeping people away from waste and disease in order to maintain health
Seamstress	Person who makes clothes by sewing
Security	Protecting people from harm
Sensor	Device that detects changes in the environment
Skyline	Shapes a group of city buildings make against the sky
Solar	Related to the sun such as solar energy which is energy created by sunshine
Stakeholder	Person who is interested in the result of an effort or who will be affected by it
Stores	Buildings in a city where products or services are provided
Structure	Item made by people such as buildings and bridges
Suburban	Area with houses between a city downtown and the countryside
Sustainability	Ability to meet needs today while preserving the needs of the future
Sustainable	Able to be maintained for many years
System	Group of items that work together to produce a result

T

Team	Group of people who work together for a common result
Technology	Something developed using scientific knowledge
Tidal	Related to the rise and fall of ocean water. It can be used to generate energy
Transportation	Ways that people can move from place to place
Turbines	Large blades that rotate and convert the wind into energy

U

Urban	Areas with a high number of people living close to each other such as a city
Urban Challenge	Goal given to teams to find solutions to a problem in a city

W

Warehouse	Large building where products are stored
Waste	Materials that are unwanted or unusable
Waves	Movement of surface water that can also be converted into energy
Wind (power)	Use of turbines to produce energy
Workability	Degree to which good jobs are created and available

About the Authors

Dr. Jonathan Reichental loves **technology**, but more importantly he loves **technology** that improves lives and makes people happier. He is fascinated by cities too. He recognizes that cities have lots of problems to solve and enjoys helping to create and deliver bold solutions, often using **technology**. Jonathan is also a passionate educator who teaches at several universities and creates online learning videos. He wrote the best seller, Smart Cities for Dummies. You can contact him and also learn more about his interests and work here: www.reichental.com.

Brett Hoffstadt is a lifelong learner and creator who loves combining his technical, analytical, and creative passions into projects that make a unique and meaningful impact in the world. His technical background includes aeronautical **engineering**, project **management**, and systems **engineering**. His creative energies have been expressed into outlets such as original music compositions, inventions, corporate **innovation** programs, and nonfiction books. His first book, published in 2014, was How To Be a Rocket Scientist. He recently published Goodnight Moon Base which is available online and in bookstores.

Acknowledgements

We are so grateful for the opportunities to work on projects that matter. It's a special gift and privilege we've both been given.

In creating this book, we want to thank a number of people and organizations who helped to make it possible.

Thanks to Freepik.com and brgfx (https://www.freepik.com/brgfx) for all the character images and graphics used within the book.

We're grateful to Google for providing so many free online tools. We used many of them in producing this book.

We're thankful for the brilliant teachers who were reviewers of the content of this book that was taken from *Exploring Smart Cities Activity Book for Kids*: Deborah Foehrkolb Belcourt, Robert Foster, Kelly Gary Kostakis, and Bernadette Brandsma.

(By the way, we love and appreciate all teachers.)

To anyone who gave us advice, did work, or just listened when we needed someone to listen, we thank you.

Finally, our biggest thanks is for YOU, the explorer! We thank you both for your interest in the important topic of cities and for being part of our journey.

Jonathan & Brett

Additional Resources

Other Books in The Steam-Powered Kids Series:
Exploring Smart Cities Activity Book for Kids
www.smartcitybook.com/kids

For grown-ups and big kids, check out Dr. Reichental's comprehensive and fun book for building better cities, *Smart Cities for Dummies*, here: www.smartcitybook.com

You can also find lots of videos and articles on smart cities and other topics at his website: www.reichental.com

Brett has produced many other activity books. Be sure to check them out and learn more here: www.HowToBeaRocketScientist.com

Copyright © 2021
Dr. Jonathan Reichental
& Brett Hoffstadt
All Rights Reserved

This book is a product of
Innovation Fountain
www.innovationfountain.com

For bulk and school orders, contact us through
www.smartcitybook.com/kids

Made with love in California, USA

www.ingramcontent.com/pod-product-compliance
Lightning Source LLC
Chambersburg PA
CBHW081329040426
42453CB00013B/2347